A Closer Look at
GRASSLANDS

A CLOSER LOOK BOOK

Published in the United States by Gloucester Press in 1979

All rights reserved

Originated and designed by
David Cook and Associates
and produced by
The Archon Press Ltd
70 Old Compton Street
London W1V 5PA

First published in
Great Britain 1979 by
Hamish Hamilton
Children's Books Ltd
Garden House
57–59 Long Acre
London WC2E 9JL

The author wishes to acknowledge the assistance received from Mrs Joyce Pope of the British Museum (Natural History), London, during the preparation of this book

Printed in Great Britain by
W. S. Cowell Ltd
Butter Market, Ipswich

Library of Congress Cataloging in Publication Data

Horton, Catherine.
   A closer look at grasslands.

   (A Closer look book)
   Includes index.
   SUMMARY: Discusses the plant and animal life of the world's grasslands, which make up almost a quarter of the world's land area.
   1. Grassland ecology—Juvenile literature. [1. Grassland ecology. 2. Ecology] I. Wilson, Maurice Charles John, 1914– II. Title.
QH541.5.P7H67   1979   574.5′264   79-14371
ISBN 0-531-03411-9

A closer Look at GRASSLANDS

Catherine Horton

Illustrated by Maurice Wilson

574.5
HOR

Gloucester Press · New York · Toronto · 1979

Grasslands and savannah

The rainy season
On tropical savannahs the rainy season is a period of sudden, dramatic change. At the end of the dry season the land looks like a dust bowl. After the first rainfall it changes overnight into lush green meadows as the grass begins to grow.

Almost a quarter of the world's land area is made up of grasslands—vast open spaces where almost nothing grows except grass. Grasslands in tropical regions near the equator are called savannahs. Temperate grasslands go by many different names: those in North America are the prairies; in Africa, the veldts; in Europe and Asia, the steppes; in South America, the pampas.

These vast seas of grass flourish in areas where there is too little rain for forests. The grasslands of the temperate zone are found in regions where the summers are usually dry and warm and the winters are very cold. These conditions discourage the growth of trees and woody plants, but tough grasses flourish.

Tropical grasslands, or savannahs, usually have one or two rainy seasons followed by long periods of drought. Most savannahs have trees or tall bushes dotted here and there over the landscape or along the banks of streams, but the principal vegetation is grass.

Some grasslands are ancient. The great African savannahs, for example, probably evolved about 65 million years ago. Other grasslands are newer, and many have been made by humans. In large areas of India and western Europe, grasslands were created when the forests were burned or cut down to clear the land for farming.

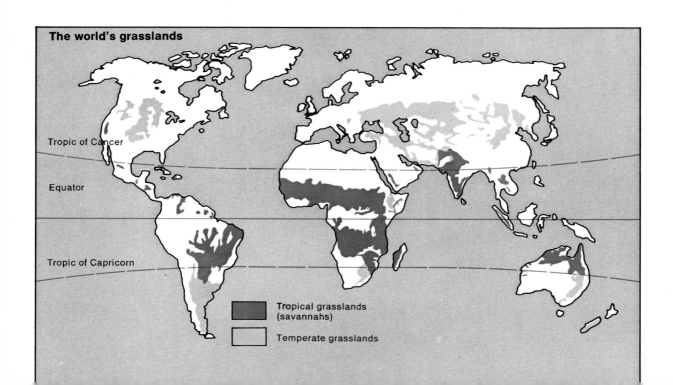

The South American pampas
The gently rolling pampas of South America are windswept grasslands where the temperature varies sharply. The pampas are one of the largest temperate grasslands in the world, covering 300,000 square miles (777,000 sq km) of southern Brazil, Uruguay, and Argentina.

The Australian grasslands
Long droughts are frequent in the Australian grasslands, but when there is enough rain the grasses thrive. Much of the natural vegetation, such as kangaroo grass, has been replaced by European species, and many of the eucalyptus trees that once dotted the plain have disappeared.

The Eurasian steppes
The steppes of Europe and Asia are the largest of the world's continental plains. They stretch all the way from eastern Europe to Siberia. Some parts of the Asian steppe are in their natural state, but most of the European plains have been cultivated or used for grazing.

The North American prairie
The grasslands of the United States and Canada are seemingly endless plains unbroken by trees. They stretch across the middle of North America. Variations in climate and rainfall result in tall-grass prairies in the eastern plains and short-grass prairies in the drier western parts.

The African savannah
Savannahs have a large variety of plant life, and on the African savannah the grasses range from low-growing species to head-high elephant grass. Scattered trees and shrubs grow in many areas. The most common trees are the acacias, shown in the picture at the left.

The grasses

Grassland plants must be able to survive in spite of irregular rainfall and long periods with no rain at all. They must also be able to withstand constant cropping and trampling by animals that feed on grass. Fire is another hazard in dry areas, and grass must be able to live in spite of it. Grass is ideally suited to survive under these conditions. Over millions of years it has become one of the most successful forms of plant life.

Most grasslands have many different species of grass as well as other plants. The grass family, Graminae, has about 8,000 species. Some are annual grasses—grasses that complete their life cycle in one growing season and then die. Annual grasses thrive in places where most other plants cannot grow. Their deep roots search out water and break down soil particles. When annual grasses die, they add valuable materials to the soil.

Perennial species are grasses that live for several years. They can store food in their roots. This enables them to survive through fire or to be dormant through periods of drought. The dead leaves and stems of these grasses protect young shoots during cold or dry seasons. When the rains come, the plants are ready to grow again.

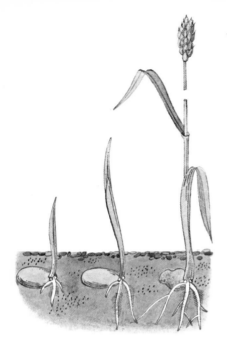

A grass plant
Grasses belong to a class of plants called monocotyledons—those that have only one seed-leaf. Grass has hollow stems, slender roots, and long, narrow leaves with parallel veins. The flower heads are made up of many small stems, each bearing clusters of tiny blossoms.

Grassland fires
Each red oat grass seed has a long spiraling shaft (*below*) which alternately expands and contracts, burying the seed where it is protected from heat and flame. Fires may provide a feast for marabou storks and secretary birds (*right*), which catch small rodents and reptiles fleeing from the flames.

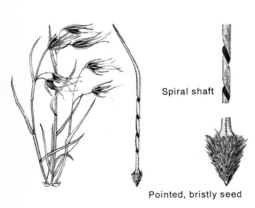

Spiral shaft

Pointed, bristly seed

Turkey buzzard

Secretary bird

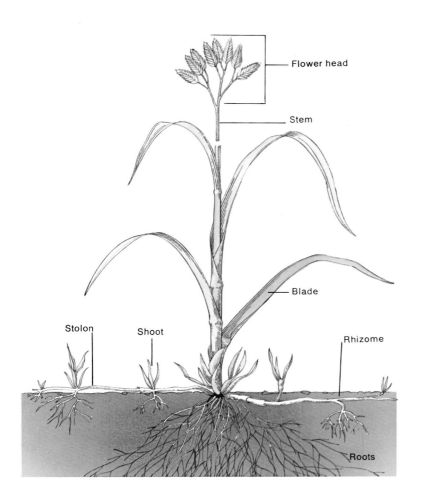

Regeneration

Grasses grow from the base of the stem, not from the tip. They regenerate, or sprout again, after the tops have been eaten by grazing animals. Many grasses spread by putting out new shoots. Stolons are stems that grow along the ground; rhizomes are underground stems. Both produce rapidly growing new shoots called tillers. Not every grass plant has all these devices. The diagram at the left does not portray a real plant. It is meant to show all the ways grasses can regenerate.

Sowing seed

Most grasslands are windswept areas, and the seed of many grasses is carried by the wind. Some seeds are able to survive in the digestive systems of animals. Others can lie dormant for long periods and then sprout when conditions are favorable for growth.

Grass as food

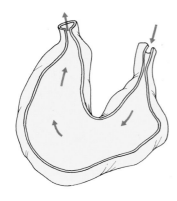

The stomach of a horse
The horse has a straightforward digestive system. Its stomach is able to deal with large quantities of coarse food such as grass which has been chewed to a pulp.

Every part of a grass plant—roots, stem, leaves, and flower heads—provides food for some kind of animal. The grasslands can feed a great variety of plant-eaters, or herbivores, from insects, birds, and small animals to large hoofed mammals, or ungulates.

Grass is difficult to digest. It contains valuable proteins, minerals, and other nutrients, but they are locked inside the plant cell, surrounded by a tough outer wall made of cellulose.

Animals with simple stomachs, such as the horse, digest their food rapidly and must first break down the plant tissue by chewing. Horses have strong teeth for grinding the grass down to a pulpy substance before swallowing it. The tissues of many grasses contain a hard substance called silica. If horses' teeth were not open-rooted (meaning that they continue growing as long as the horse lives), they would soon be worn away by the silica.

Grazers of the steppes
Wild asses have simple stomachs, as do horses. Hares re-eat grass pellets which they have already digested once. Susliks (Eurasian ground squirrels) are among the many seed-eating animals of the grasslands.

Most other grass-eaters are ruminants. These are animals with complex stomachs. They take a long time to digest their food. Ruminants return partly digested grass from the stomach to the mouth and chew it a second time. This is called chewing the cud. The food passes back and forth from mouth to stomach and from one stomach chamber to another. Rumination is a slow process, but it has certain advantages. Animals are most likely to be attacked while they are feeding. Those like horses who must chew their food on the spot are exposed to danger for a longer period of time. Ruminants are in less danger because grazing and chewing are two different activities; chewing can be done later in a safer place.

A third way of digesting grass is the one used by rabbits and hares. They re-eat pellets of grass after it has passed through their bodies the first time. This process of digestion is called refection.

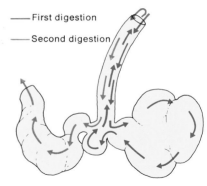

The stomach of a ruminant
Ruminants have four-chambered stomachs. The first two chambers deal with partly chewed food. This is returned to the mouth (regurgitated) and chewed again. The pulp then goes to a third chamber and finally to the abomasum, or true stomach.

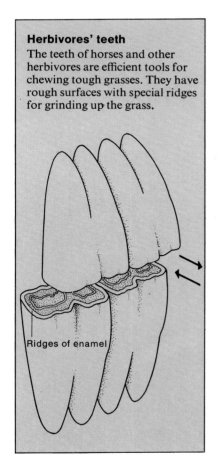

Herbivores' teeth
The teeth of horses and other herbivores are efficient tools for chewing tough grasses. They have rough surfaces with special ridges for grinding up the grass.

Ridges of enamel

Ruminants of Eurasia
Saigas (goat antelopes of Eurasia) have complex stomachs. Like other ruminants, they swallow partially chewed food and later chew the cud—perhaps in a spot where they are safer from predators.

Golden eagle

Saigas

Survival

Most animals engage in a constant struggle to find food and escape from their enemies. The way in which the laws of survival operate could be seen in a natural grassland, such as the North American prairie before the arrival of the European settlers. Until the nineteenth century, the prairies were a vast grass-covered wilderness where enormous herds of bison and pronghorn antelope roamed. Packs of hungry wolves followed the herds waiting for a chance to make a kill. This prairie scene illustrates what ecologists call a food chain—the way in which energy from the sun is changed by plants into food for plant-eating animals (herbivores); in their turn, the plant-eating animals are eaten by meat-eaters (carnivores).

Grass is called a primary producer; grass plants are able to convert the sun's energy and store it in the form of chemical compounds. Then primary consumers, like the bison and the antelope, feed on the grass. Secondary

The vanishing prairie
Although the American prairies are still vast, they are much less so than they used to be. The advance of modern civilization over the past 150 years has taken its toll. The huge herds of bison and pronghorn antelopes are gone. Small herds of these animals are being reintroduced today.

consumers, predators who live by preying upon other animals, in turn eat the herbivores.

Herbivores are not easy prey, so finding food can be difficult for carnivores. Pronghorn antelopes, for example, are among the fastest runners on earth. A healthy antelope can easily outdistance a wolf. Living in herds is further protection as it is difficult to attack animals when they are grouped together. Moreover, the herd is often warned of danger in time to escape. Bison live in herds, but their chief defense is their size. A male bison may weigh 3,000 pounds (about 1,350 kg).

Wolves pick out and attack the feeble members of a herd—the sick, the young, and the old. As a result, the strongest and healthiest members of the herd survive to reproduce. Killing off a certain number of the young prevents overcrowding and overgrazing. In the end, the herd profits from the eating habits of predators.

Extinct wolves
Until the mid–1800s, wolves were the chief predators of the American prairies. The "white wolf" (*below*) had become extinct by the twentieth century. It was destroyed by settlers, who saw the wolf only as a dangerous pest—not as an important part of a living community.

Life underground

In the great grasslands there is life underground as well as on the surface. Small animals burrow in the earth where they escape from enemies and avoid fire and extremely hot or cold weather. In dry grasslands the temperature may vary from sizzling hot at noon to below freezing at night.

Most grassland burrowers belong to the order of Rodentia, the rodents. Their sharp, constantly growing incisor teeth are perfect for gnawing the seed, husks, roots, and leaves they eat. Some of these rodents, like the North American pocket gophers, are solitary creatures. Others, like the prairie dog and the bobac marmot of Europe and Asia, live in colonies. They dig large underground quarters, 10 feet (about 3 m) or more below the surface. The burrows often hold enormous numbers of animals. One prairie dog colony discovered in the early

Some small rodents
Many different species of mice and voles live in dry grasslands all over the world. Lemmings are found on the Asian steppe and the Arctic tundra.

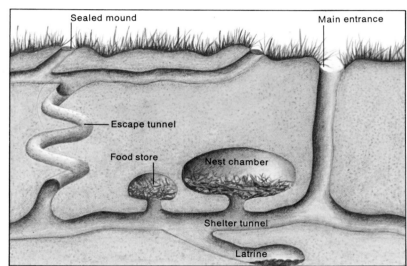

Gophers and prairie dogs
The valley gopher (*above*) is a North American ground squirrel. It builds an elaborate burrow (*right*) where it leads a solitary life. Below is a prairie dog "city," a vast network of burrows whose entrances are marked by small craters.

1900s covered about 24,000 square miles (about 61,000 sq km) and is said to have housed 400 million animals!

Like other animals, burrowers play an important part in the grassland community. By tunneling through the earth, they mix and aerate the soil, stimulating the growth of plants. They themselves are the food of many predators (hunters that prey on other animals), and their burrows provide homes for other ground-dwellers.

The burrowing owl often nests in deserted tunnels of prairie dog colonies. This owl hunts during the day for mice, rats, and birds and also preys on young prairie dogs. In turn, the eggs and young of the owl are eaten by rodents. The owl tries to intimidate these hunters by imitating the warning signal of the rattlesnake, another frequent visitor to the burrows. The rattlesnake moves through the tunnels searching for food, or resting away from the heat of the sun.

Small predators
Owls and hawks have an advantage over ground-dwelling predators—they can spot their prey from the air. Badgers use their sharp claws to dig their victims out of the earth. Weasels, skunks, and snakes (like the prairie rattlesnake at the left) also prey upon rodents.

Australia and South America

Mara

The South American pampas
The viscacha (*below*) and the mara (*above*) are pampas rodents. Unlike other rodents, the mara has long legs and uses them to bound across the plains. The flightless rhea (*below*) has some habits in common with the larger ostrich of Africa. Both birds live in flocks and eat all kinds of foods.

Areas like these that are isolated geographically often produce native species found nowhere else. Some interesting marsupial mammals native to Australia are kangaroos, koalas, and wombats. (Marsupials give birth to small, helpless young which are carried and nursed in a pouch on the mother's abdomen until they are mature enough to emerge.)

Some special animals of the South American pampas include the armadillo with its armor of bony plates, the vizcacha (a rodent), and the Patagonian cavy, or mara.

Although the grasslands of South America and Australia are much like those in other parts of the world, particular species have evolved due to the isolation of these continents. Yet many isolated species are strikingly similar in behavior and appearance to a different species, found elsewhere. This seeming coincidence is called convergent evolution. It comes about because the different species have had to adapt to similar conditions. Some good examples are the long-legged, flightless birds of Africa, South America, and Australia—the ostrich, the rhea, and the emu.

Australian grazers
Some fifty species of kangaroo live on the Australian grasslands. The great red (*below*) can weigh up to 220 pounds (about 100 kg). Smaller species of kangaroos are called wallabies. Kangaroos thrive in a dry climate and eat many kinds of plants.

The last great assemblies

At one time large herds of hoofed mammals, or ungulates, grazed on the open plains of most of the world's grasslands. Now few large herds remain except on the African savannah. Nowhere else in the world are there such numbers or so many species of large grazing animals.

When the rains come to East Africa, great herds of these animals make their annual trek from the river valleys, where they have spent the dry season, to the lush pastures of the open plains. Thousands of zebras and wildebeests (gnus) travel along migration routes which have been used by these animals for centuries. The herds form a moving mass that is one of the last great wildlife spectacles on earth.

While the rains fall and food is abundant, these animals share the plains with thousands of other grazers.

One plant serves many
Zebras, wildebeests (gnus), and Thomson's gazelles each eat a different part of the same plant. Zebras eat the stems, wildebeests the leaves, and gazelles the tender new growth.

Many, like the wildebeest, are antelopes. The males—and often the females—have hollow, cylindrical ringed horns that are never shed. Among the antelopes that live in open grassland are Thomson's gazelles, elands, topis, and hartebeests. Both the antelope and the zebra have long, muscular legs for traveling long distances over rough ground and for running to escape predators.

Such a great assembly of animals can live together only because most of them have their own distinct feeding habits. Some eat the coarse outer parts of grass; others the tender new shoots which sprout after the tougher parts have been cropped. The huge eland feeds on the leaves of wild flowers that grow among the grasses. As the wild flowers are rather scarce, elands are found to be more widely scattered than other antelopes.

The springhaas
The springhaas, or Cape jumping hare, lives in large colonies on the open plains of Africa. It feeds on roots and bulbs which it digs up with its strong back legs.

Wildebeest (gnu)

Plant-eaters of the Savannah

The African savannahs range from extremely dry areas to regions of swamp, lake, and river, to woodland. Because of this variety, the savannahs can support a larger number and variety of herbivores, or plant-eaters, than any other area in the world. The diagram (*right*) shows where different herbivores live. Water-loving antelopes such as the reedbuck, Uganda kob, lechwe, and waterbuck are never very far away from rivers, lakes, or swamps. Grant's gazelles and oryxes can stay in dry areas for long periods, getting the moisture they need from plants.

Some herbivores eat leaves and twigs as well as grass. They are often found in the border country between open grassland and wooded savannah, where they take advantage of the available variety of food. Usually animals that live in woodland and bush are adapted for reaching into trees to gather food. Many savannah trees have thorns to protect them from the browsing animals. Giraffes and browsing antelopes have long pointed snouts for gathering leaves and stems while avoiding the thorns.

Trees and shrubs provide food at every level: long-necked giraffes can reach the highest leaves; gerenuks stand on their hind legs to eat from the middle branches; and the tiny dik-dik browses on low-growing shrubs.

Animals of water and marsh
The waterbuck (*above*) is a powerful swimmer well adapted to marshy ground, reed beds, and flooded grassland. It feeds in tall-grass areas as well. Hippos (*below*) are also found in wet areas and are able to remain underwater for ten to twelve minutes.

Herbivore habitats

Open grassland... giving way to wooded savannah... and dense bush

Black rhinos
The black rhino is a browser, feeding on the twigs and leaves of low-growing shrubs. It can eat much coarser vegetation than its grazing relative, the white rhino.

The giant plant-eaters

The world's largest land animals are found on the African savannah. These huge animals are all vegetarians (herbivores). The African elephants, largest of them all, can adapt to many different habitats and eat a variety of foods. They live in open grassland, in bush and forest, near lakes, and in mountainous country. They feed on tall grasses, bark, leaves, fruit, roots, and berries. Elephants often live in very large herds. The herds are made up of smaller family groups of four to twenty elephants, led by the older females.

White rhinoceroses are grazers weighing more than three tons. Once white rhinos were found all over Africa, but now they are one of the continent's rarest animals. Their herds are made up of very small family groups of one male, one or two females, and several young.

Their smaller and more solitary relatives, the black rhinos, are more numerous and more widely distributed. The black rhino is a browser with a pointed, finger-shaped upper lip for grasping leaves and branches.

Neither rhinos nor elephants are much bothered by predators. They are protected by their size and by their formidable horns and tusks. Their only real enemies are humans, who have hunted them mercilessly and have succeeded in making some species nearly extinct.

White rhino
The "white" rhino's color is actually dark gray. Its name comes from the Boer word *weit*, meaning "wide," used to describe the enormous width of the rhino's mouth.

African elephants
Elephants are big eaters. They must forage for vast quantities of vegetable food each day, pulling up tufts of grass and stripping leaves, twigs, and bark from trees. As a result, elephant herds have turned many forested areas into savannahs by destroying the trees.

Predators

The great predators of grassland and savannah belong mainly to two families: the cats, or Felidae, and the dogs, or Canidae. In Africa the big cats are the lion, the cheetah, and the leopard. They are large hunters with powerful fangs, keen sight, and remarkable reflexes. With the exception of lions, big cats are solitary hunters who often attack from ambush. Leopards may spend hours stalking their prey. Cheetahs rely on their great sprinting speed to run down their prey. Only lions work in teams, with the females doing most of the work and a single lioness usually making the kill. A big cat kills its victim quickly by breaking its back or more slowly by seizing the animal and suffocating it.

The Canidae are smaller, slender animals with very strong jaws, sharp teeth, and an excellent sense of smell. The African wild dogs live and hunt in packs. A pack may have from six to twenty members. Wild dogs are long-distance runners with tremendous staying power. While chasing their prey, they tear at its body until the animal collapses, exhausted. The meat is shared by the pack.

The jackal and hyena (of the Hyaenidae family) are other important predators. They kill a good deal of their own food and also feed on the kills of other predators.

The leopard
The leopard is an excellent climber, and in forested areas it will ambush its prey from the branch of a tree. After the kill it sometimes drags its victim up into the tree, where it can feed undisturbed.

Whose kill?
Hyenas are often thought of as cowardly scavengers. Actually they kill much of their own food. Lions, who are lazy hunters, sometimes move in and drive the hyena away from its kill.

Wild dogs
Discipline and teamwork are required of animals that hunt in packs. Wild dogs (*right*) work closely with one another once they have picked out a victim.

The cheetah
This big cat usually stalks its prey for thirty minutes or more before bursting into a sprint (*below*) and overtaking its victim. The cheetah runs at very high speed for short distances but lacks endurance.

Scavengers

Spotted hyena

Striped hyena

When predators like lions and leopards have finished gorging themselves on the carcass of a dead animal, the scavengers move in to finish off the remains. Vultures and marabou storks look for food during the day; hyenas and jackals feed at night. When the larger scavengers have finished, smaller creatures—ants, rats, crows, and ravens—feed on the meat. Finally, bacteria break down bones, and fungi attack horns; every part of a dead animal is used by some living thing. Scavengers perform a valuable service by getting rid of dead bodies which might otherwise harbor disease-carrying organisms.

There are many vultures on the high plains of Africa, perched on dead trees or wheeling high above the ground on the lookout for food. Vultures are farsighted and have powerful wings. They glide effortlessly on rising air currents sometimes reaching a height of 6,500 feet (about 2,000 m).

Vultures of several different species often feed together in what looks like a huge free-for-all. But each species eats a different part of the carcass. Egyptian and hooded vultures eat pieces of flesh discarded by other birds. White-backed and Rupple's vultures eat internal organs, while still other varieties feed on skin, muscle, or nerve fiber.

Hyenas

Both striped and spotted hyenas live on the African savannah. The spotted hyena lives and hunts in groups called clans but sometimes forages on its own. Both species have strong teeth and jaws. They eat almost all of a carcass, including the bones.

Vultures and marabous

Vultures have curved, pointed beaks that are designed for cutting, tearing, and dismembering a carcass. The marabou stork has a long sharp bill that enables it to reach the organs far inside the body of the dead animal.

White-backed vulture

White-headed vulture

Egyptian vulture

Marabou stork

Finding the kill
Once a dead animal has been sighted, many scavengers arrive within minutes. Vultures serve as lookouts and spot the site of a kill from above. Jackals watch the sky for these great birds and follow them when they swoop down to land.

Marabou storks

White-headed vultures

Jackals

Smaller predators

Aardvark
The strange-looking aardvark lives in the African savannah. It digs out ants and termites from their nests with its powerful claws. Its long, sticky tongue can reach into the cracks and corners of nests.

Some of the smaller predatory mammals of the African savannah are pictured at the right. The caracal, or desert lynx, hunts small antelopes such as the tiny dik-dik, klipspringer, and oribi. These antelopes rely mainly on running and jumping to escape their enemies.

The black spotted serval, another of the smaller cats, hunts ground squirrels, long-tailed rats, and guinea fowl, which try to escape by hiding in the tall grass or going underground. The serval can leap high in the air after birds and catch them on the wing.

Other predators include the solitary genet, the black and white striped zorilla (which looks much like a skunk), and the bat-eared fox. Snakes, such as the common puff adder and the Egyptian and spitting cobras, kill their victims by poison. They are themselves attacked by mongooses and secretary birds. One of the most specialized of the hunters is the odd-looking aardvark, which feeds entirely on ants and termites and hunts at night.

Several species of monkeys, baboons, vervets, and patas monkeys have adapted to grasslands by living mostly on the ground. The big cats and jackals are the monkey's most dangerous enemies. Vervets are hunted by baboons and martial eagles as well.

Grassland primates
The small patas monkey (*above*) stands on its hind legs to look out for predators over the tall grass. The olive baboons (*below*) live in large troops, sharing their habitat with hyraxes and nimble-footed klipspringers.

Cultivated grasslands

Humans have been growing crops and raising domestic animals for thousands of years. Ancient Egyptians were sowing wheat 4,500 years ago, and some African tribes have kept cattle for 5,000 years. Today most of the world's temperate grasslands are under cultivation. In tropical regions, where conditions are poor for farming, grasslands are used for grazing.

Human use of the grasslands has replaced wild plants and animals with cultivated ones. Many kinds of natural grasses are replaced by a single crop; many different animals living together are replaced by only one or two domestic species. An area of cultivated grassland can feed only one-tenth as many animals as before. More space is required because cattle and sheep, unlike wild animals, all eat the same type of grass.

Most of the world's cropland is planted with highly nutritious cereal grasses such as wheat and rice. Cereals

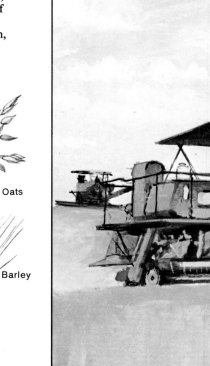

Grass crops
About 70 percent of the world's cropland is planted with cereal grasses, whose seeds provide humans with grain. Wheat, rye, and oats are the main cereals of temperate regions; they need moisture for growth, and warm, dry weather for harvesting.

are eaten by every race of mankind and in many countries are the people's main diet. Human beings are dependent in many ways on the world's great grasslands.

The question is whether people can learn to use the grasslands wisely without resorting to destructive practices. One such practice is overgrazing, which may strip the land of vegetation. The soil is then exposed to erosion by wind and water, and eventually the land becomes a desert. Another is the use of weed killers, which destroy the plants needed by many grasslands creatures for food and shelter and often leave harmful chemical residues. A third example of poor management is the use of fire to clear land for farming. Burning at the wrong season, when seeds are being produced, for example, may result in permanent loss of plant life. In the end, practices such as these can be disastrous not only for the soil itself but for all plant and animal life.

Domestic animals
Cattle, sheep, and goats provide food and clothing, but they are heavy grazers. By cropping plants to the ground, they prevent regrowth and cause water loss. If they are not well managed, grasslands sometimes become dust bowls as a result of overgrazing.

The future of grasslands

Grass is a tough and resilient plant, the first to grow again on land that has been devastated by humans or by nature. It helps bind the soil and retain moisture, lives through extreme temperatures and withering winds, and provides a basic food for many creatures. Mankind needs grass. Yet today few grasslands remain in their natural state. Some have been completely destroyed by neglect and ignorance. Those that have survived will be in danger unless they are protected.

It is not only natural grasslands which need protection. Grasslands that are used for agriculture can easily be reduced to desert if they are misused. They must

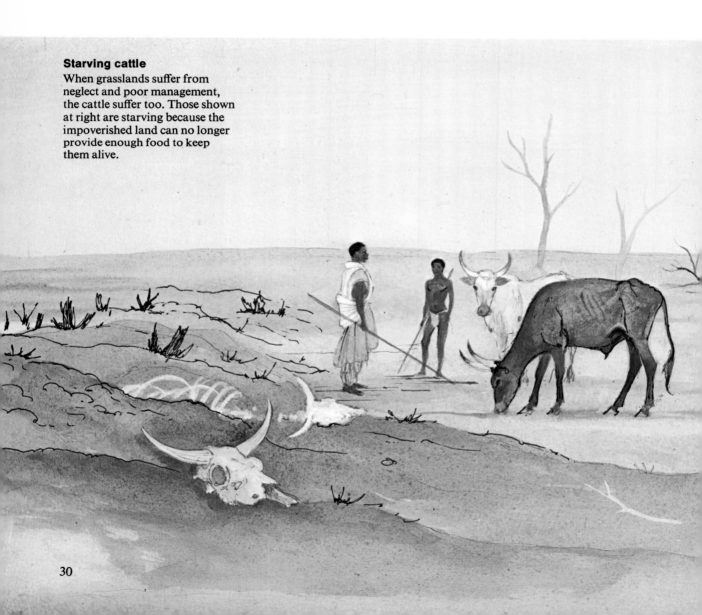

Starving cattle
When grasslands suffer from neglect and poor management, the cattle suffer too. Those shown at right are starving because the impoverished land can no longer provide enough food to keep them alive.

have good management. The land will remain fertile only if farmers conserve it by such measures as crop rotation, contour plowing, and controlled grazing.

In many countries, governments have established parks and game preserves where wild animals can roam freely in their natural environment. This has helped to protect some of the endangered species. But even these areas may be threatened. As world population grows, pressure to use land for settlements and for raising food becomes greater. People must learn to provide for their needs without harming plant and animal life if the grasslands—and the human race—are to survive.

Index

Aardvark, 26
Africa, 4, 14, 16, 17, 21, 22, 24
Agriculture, 5, 30
Annual trek, 16
Antelopes, 10, 11, 17, 18, 26
Asia, 4, 5, 12
Australia, 5, 14

Baboons, 26, 27
Birds, 8, 9, 12, 13, 14, 24, 26
Bison, 10, 11
Burrowers, 12–13

Carnivores (meat-eaters), 10, 11
Cats, 22, 26
Cereal grasses, 28
Cheetahs, 22, 23
Chewing, 8, 9
Climate, 5, 14
Contour plowing, 31
Convergent evolution, 14
Crop rotation, 31
Crops, 28

Deserts, 30
Digestion, 8–9
Dik-dik, 18, 26
Dogs, 22
Drought, 4, 5, 6

Elephants, 21
Erosion, 29
Europe, 4, 5, 12

Farming, 4, 28–29, 31
Fire, 6, 12, 29
Food, 10, 13, 16
Forests, 4

Giraffes, 18
Grass, grasses, 4–8, 10, 26, 28
Grass as food, 8–9, 10, 17, 18, 21
Grass plant, the, 6–8, 10, 30
Grass seed, 6, 7, 9
Grasslands, cultivated, 28–29, 30–31
Grazers, 8, 14, 16, 21, 29, 30

Grazing, 6, 7, 16, 28, 31
Ground squirrels, 12, 26

Hartebeests, 17
Herbivores (plant-eaters), 8–11, 18, 21
Herds, 10, 11, 16, 21
Hippopotamus, 18
Horses, 8, 9
Hyenas, 22, 24

Insects, 8, 24, 26

Jackals, 22, 24, 25, 26

Kangaroos, 14
Klipspringers, 26, 27

Leopards, 22, 24
Life cycle of grass, 6
Lions, 22, 24

Marabou storks, 6, 24
Maras, 14
Marsupials, 14
Meat-eaters (carnivores), 10, 11
Migration routes, 16
Monkeys, 26, 27

North America, 4, 5, 10

Ostrich, 14
Overgrazing, 11, 29
Owls, 12, 13

Packs, animals in, 22, 23
Pampas, 4, 5, 14
Parks and game preserves, 31
Plains, 5, 14, 16, 17, 24
Plant-eaters (herbivores), 8, 9, 10, 18, 21
Plants, 5, 8, 10, 16, 29
Prairie dogs, 12–13
Prairies, 4, 5, 10, 11
Predators, 11, 13, 17, 21, 22, 24, 26, 27

Primary consumers, 10
Primary producer, 10

Rainfall, 4, 5, 6, 16
Refection, 9
Regurgitation, 9
Rhinoceroses, 20, 21
Rodents, 9, 12, 13, 14, 24, 26
Rumination, 9

Savannahs, 4, 5, 16, 18, 21, 22, 24, 26
Scavengers, 22, 24, 25
Seasons, 4, 6, 16
Secondary consumers, 10–11
Secretary birds, 6, 26
Seed-eaters, 9
Snakes, 12, 13, 26
Soil, 6, 29, 30
South America, 4, 5, 14
Steppes, 4, 5, 12
Stomachs, simple and complex, 8, 9

Teeth, 8, 9, 12
Temperate regions, 4, 5, 28
Temperatures, 5, 12, 30
Thomson's gazelles, 16, 17
Trees, 4, 5, 18, 21
Tropical regions, 4, 28
Tundra, 12

Ungulates (hoofed mammals), 8, 16

Vegetation, 4, 5
Veldts, 4
Vultures, 24, 25

Wallabies, 14
Wheat, 28
Wild dogs, 22, 23
Wildebeests, 16, 17
Wolves, 10, 11

Zebras, 16, 17